Forces and Motion Bingo Book

A COMPLETE BINGO GAME IN A BOOK

Written By Rebecca Stark

ISBN 978-0-87386-436-7

Educational Books 'n' Bingo

Printed in the U.S.A.

FORCES & MOTION BINGO DIRECTIONS

INCLUDED:

List of Terms

Templates for Additional Terms and Clues

2 Clues per Term

30 Unique Bingo Cards

Markers

1. **Either cut apart the book or make copies of ALL the sheets. You might want to make an extra copy of the clue sheets to use for introduction and review. Keep the sheets in an envelope for easy reuse.**

2. Cut apart the call cards with terms and clues.

3. Pass out one bingo card per student. There are enough for a class of 30.

4. Pass out markers. You may cut apart the markers included in this book or use any other small items of your choice.

5. Decide whether or not you will require the entire card to be filled. Requiring the entire card to be filled provides a better review. However, if you have a short time to fill, you may prefer to have them do the just the border or some other format. Tell the class before you begin what is required.

6. There are 50 terms. Read the list before you begin. If there are any terms that have not been covered in class, you may want to read to the students the term and clues before you begin.

7. There is a blank space in the middle of each card. You can instruct the students to use it as a free space or you can write in answers to cover terms not included. Of course, in this case you would create your own clues. (Templates provided.)

8. Shuffle the cards and place them in a pile. Two or three clues are provided for each term. If you plan to play the game with the same group more than once, you might want to choose a different clue for each game. If not, you may choose to use more than one clue.

9. Be sure to keep the cards you have used for the present game in a separate pile. When a student calls, "Bingo," he or she will have to verify that the correct answers are on his or her card AND that the markers were placed in response to the proper questions. Pull out the cards that are on the student's card keeping them in the order they were used in the game. Read each clue as it was given and ask the student to identify the correct answer from his or her card.

10. If the student has the correct answers on the card AND has shown that they were marked in response to the *correct questions,* then that student is the winner and the game is over. If the student does not have the correct answers on the card OR he or she marked the answers in response to *the wrong questions,* then the game continues until there is a proper winner.

11. If you want to play again, reshuffle the cards and begin again.

Have fun!

TERMS

acceleration	magnet(s)
balanced	mass
compound machine	matter
convert(ed)	momentum
efficiency (efficient)	motion
electromagnetism	newton
energy	Sir Isaac Newton
float	physics
force	potential energy
free fall	power
friction	pull
fulcrum	pulley
Galileo	push
gear	resistance
gravity	scientific method
hypothesis	screw
inclined plane	simple machines
inertia	speed
joule	suction
kinetic energy	tension
law	velocity
lever(s)	wedge
load	weight
lubricant	wheel and axle
machine	work

Additional Terms

Choose as many topics as you would like and write them in the squares.
Repeat each as desired. Cut out the squares and randomly
distribute them to the class.
Instruct the students to place the square on the center space of their card.

Forces and Motion Bingo

© **Barbara M. Peller**

Clues for Additional Terms

Write two or three clues for each new topic.

1.	1.
2.	2.
3.	3.
1.	1.
2.	2.
3.	3.
1.	1.
2.	2.
3.	3.

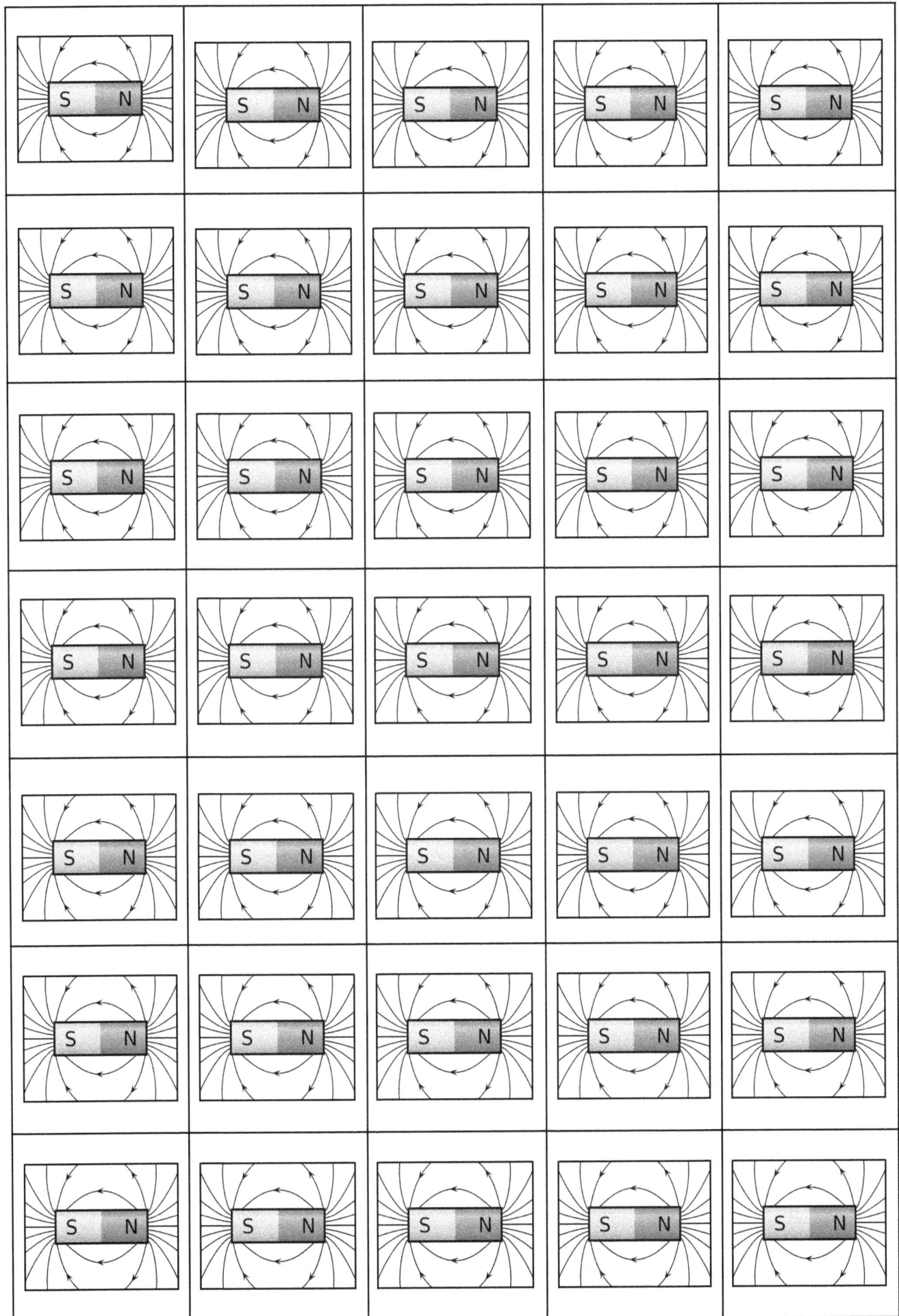

acceleration 1. ___ is the rate of change in velocity. 2. Newton's second law of motion states that the ___ of an object depends on its mass and on the size and direction of the force acting on it.	**balanced** 1. Forces that are equal in size and opposite in direction are said to be ___. The net force is zero. 2. ___ forces exert the same amount of force on an object; therefore, there is no change in the object's motion.
compound machine 1. A combination of two or more simple machines working together is a ___. 2. Scissors use levers to force wedges onto an object; therefore, scissors are a ___.	**convert(ed)** 1. Energy can be ___ from one form to another. 2. Solar panels can capture light energy from the sun and ___ it to electricity.
efficiency (efficient) 1. ___ is the relationship between the energy needed and the work accomplished. 2. It is more ___ to move a large load of dirt onto a truck with a backhoe than with a shovel.	**electromagnetism** 1. ___ is magnetism produced by an electric current. 2. The use of a copper wire carrying an electric current to magnetize pieces of iron or steel is an example of ___.
energy 1. ___allows us to do work or to move things. 2. Heat, light, sound, electrical, wind, and movement are some forms of ___.	**float** 1. If an object is less dense than the fluid it is in, the object will ___ in the fluid. 2. If an object is more dense than the fluid it is in, the object will not ___ in the fluid.
force 1. A ___ is a push or a pull that causes an object to move, stop or change its direction. 2. Newton's third law of motion tells us that for every ___, or action, there is an equal and opposite ___, or reaction. Forces and Motion Bingo	**free fall** 1. The action of gravity on an object without any device to provide resistance results in a state called ___. 2. The parachute is a drag-producing device; it prevents the falling body from being in a state of ___.

friction 1. A force that resists motion when two surfaces rub together is called ___. 2. ___ slows down or stops a moving object that is touching another object or surface.	**fulcrum** 1. The support or pivot about which a lever turns is called the ___. 2. An oar is a lever. The oarlock, or pivot point, is the ___ of the lever.
Galileo 1. This scientist's study of forces helped Isaac Newton in the development of his laws of motion. 2. This scientist was born in 1564. He died in 1642, the year Isaac Newton was born.	**gear** 1. A ___ is a toothed wheel. 2. When a ___ meshes with another toothed part, it transmits motion and may change the speed or direction of the applied force.
gravity 1. The force of attraction between objects is called ___. 2. ___ is the force of attraction that causes objects to fall toward the center of the earth.	**hypothesis** 1. A ___ is a tentative assumption based upon observations. 2. A ___ can be supported or refuted through experimentation or more observation.
inclined plane 1. An ___ is a flat surface whose endpoints are at different heights. Unlike a wedge, this kind of simple machine stays still. 2. A ramp is a type of ___.	**inertia** 1. ___ is the property of matter that keeps it moving in a straight line or at rest. 2. ___ causes your body to move forward when you are riding in a car and the driver slams on the brakes .
joule 1. A ___ is a unit of electrical energy. 2. This unit of electrical energy is equal to the work done by a force of one newton acting through a distance of one meter. Forces and Motion Bingo	**kinetic energy** 1. ___ is energy in motion or use. 2. An object waiting to be dropped has potential energy, but an object dropped from a balcony has ___. © Barbara M. Peller

law	**lever(s)**
1. A ___ is a scientific statement that generalizes a body of observations. At the time a ___ is made, no exceptions have been found to contradict it. 2. You could use the ___ of gravity to predict what will happen if you drop an object.	1. A ___ is a simple machine consisting of a rigid bar that can pivot on a fulcrum. 2. There are three classes of ___. A seesaw, a can-opener, and a fishing rod are examples of each class.
load	**lubricant**
1. The weight to be carried or moved is the ___. 2. In a first-class lever the fulcrum is between the effort & the ___. In a second-class lever the ___ is between the fulcrum & the effort. In a third-class lever the effort is between the fulcrum & the ___.	1. A ___ is a substance used to reduce the friction between two solid surfaces. 2. Sometimes a ___, such as grease or oil, is added to moving parts to reduce friction.
machine	**magnet(s)**
1. A ___ is a mechanical device that transmits, modifies, or changes the direction of force. 2. A ___ makes work easier by changing the size or the direction of a force.	1. Like poles of two ___ repel; unlike poles of two ___ attract. 2. A ___ attracts iron and produces a magnetic field.
mass	**matter**
1. The amount of matter in an object is its ___. 2. ___ is a measure of how much matter an object has. Weight is a measure of how strongly gravity pulls on that matter.	1. ___ is anything that has mass and takes up space. 2. Three states of ___ are liquid, solid and gas.
momentum	**motion**
1. ___ is how hard it is to slow down or stop a moving object. 2. The larger an object is and the faster it is going, the more __it has.	1. ___ is movement of an object from one place to another. A force is needed to put something in ___. 2. The following are types of ___: circular, rolling, vibrating, and sliding.

Forces and Motion Bingo

newton 1. This is the international unit of force. 2. This unit of measurement was named for the scientist known for his laws of motion.	**Sir Isaac Newton** 1. His first law of ___ states that unless a force causes it to do otherwise, an object at rest will remain at rest and an object in motion will stay in motion. It is also called the Law of Inertia. 2. His third law of motion tells us that for every force, or action, there is an equal and opposite force, or reaction.
physics 1. ___ is the study of matter and its motion. 2. ___ is the science of matter and energy and of the interactions between them.	**potential energy** 1. ___ is stored energy. It is energy waiting to be used. 2. An object dropped from a balcony has kinetic energy. An object waiting to be dropped has ___.
power 1. ___ is the amount of work done for each period of time. 2. ___ is the rate at which work is done.	**pull** 1. A force that causes motion toward the source of the force is called a ___. 2. A force may be defined as a push or a ___.
pulley 1. A ___ is a simple machine that uses grooved wheels and a rope to raise, lower or move a load. 2. A block and tackle is a machine formed by one fixed ___ and one movable ___ working together.	**push** 1. A force that causes motion away from the source of the force is called a ___. 2. A force may be defined as a pull or a ___.
resistance 1. A synonym for ___ is "drag." 2. A parachute produces ___, or drag.	**scientific method** 1. The ___ is a procedure used to test a hypothesis by making predictions about the outcome of an experiment. 2. The steps in the ___ are state the problem; form a hypothesis; make predictions; perform an experiment; and interpret the results.

screw	**simple machines**
1. This simple machine is an inclined plane wrapped around a shaft. 2. The bottom of a light bulb is this kind of simple machine.	1. There are six types of ___: inclined plane, screw, pulley, lever, wedge, and wheel and axle. 2. A combination of two or more ___ working together is called a compound machine.
speed	**suction**
1. ___ describes how fast something is moving. 2. ___ is a measure of the distance an object moves in a given amount of time.	1. ___ is a force that causes a fluid or solid to be drawn into an interior space or to adhere to a surface. 2. A vacuum cleaner uses ___ to clean carpets.
tension	**velocity**
1. ___ is a force tending to stretch or elongate an object. It is also the measurement of such a force. 2. When the surface of a liquid in contact with air or other gas acts like a thin elastic sheet, it is called surface ___.	1. The rate of change in ___ of a moving body is called acceleration. 2. An object's speed in a particular direction is its ___.
wedge	**weight**
1. This simple machine is thick at one edge and tapered to a thin edge at the other. It is used for insertion in a narrow crevice. 2. A ___ can be used to separate objects or parts of an object, to lift an object, or to hold an object in place. An axe is one.	1. ___ is a measure of the heaviness of an object. 2. ___ is a measure of how strongly gravity pulls on matter. Mass is a measure of how much matter an object has.
wheel and axle	**work**
1. A ___ is a simple machine consisting of an axle to which a wheel is fastened. 2. A ___ is actually a lever that rotates in a circle around a fulcrum. A bicycle wheel is an example of this simple machine.	1. The force used to move something is called ___. 2. Machines do ___. The formula for ___ is Force x Distance = ___.

Forces and Motion Bingo

inertia	compound machine	load	wedge	simple machines
force	joule	tension	matter	law
resistance	physics		lever(s)	motion
velocity	balanced	inclined plane	work	lubricant
machine	wheel and axle	friction	acceleration	kinetic energy

Forces and Motion Bingo

wedge	screw	magnets	Sir Isaac Newton	machine
lubricant	gravity	float	balanced	push
power	wheel and axle		fulcrum	inclined plane
matter	pull	physics	weight	law
kinetic energy	tension	friction	force	acceleration

Forces and Motion Bingo

wedge	inclined plane	matter	work	resistance
wheel and axle	compound machine	electro-magnetism	joule	momentum
balanced	tension		pulley	convert(ed)
physics	power	machine	gravity	magnets
acceleration	friction	force	weight	load

Forces and Motion Bingo

resistance	work	matter	inclined plane	wedge
machine	gravity	friction	power	physics
inertia	weight	force	motion	acceleration

Forces and Motion Bingo

physics	pulley	load	friction	machine
mass	gravity	joule	Sir Isaac Newton	resistance
lever(s)	float		simple machines	work
inclined plane	hypothesis	tension	force	electro-magnetism
acceleration	kinetic energy	newton	gear	motion

Forces and Motion Bingo

kinetic energy	simple machines	balanced	float	friction
mass	inclined plane	electro-magnetism	physics	Galileo
screw	motion		compound machine	load
law	pulley	inertia	weight	gear
matter	force	potential energy	fulcrum	lever(s)

Forces and Motion Bingo

convert(ed)	pulley	magnets	screw	motion
work	balanced	gear	joule	resistance
Sir Isaac Newton	electro-magnetism		float	fulcrum
force	machine	weight	newton	lever(s)
lubricant	inclined plane	inertia	potential energy	load

Forces and Motion Bingo

inertia	pulley	scientific method	Galileo	matter
lubricant	load	wheel and axle	compound machine	resistance
magnets	work		fulcrum	efficiency (efficient)
physics	gravity	mass	wedge	power
friction	force	weight	newton	convert(ed)

Forces and Motion Bingo

lever(s)	pulley	energy	work	efficiency (efficient)
mass	screw	Sir Isaac Newton	load	simple machines
resistance	push		motion	float
acceleration	physics	wedge	gear	gravity
tension	force	newton	balanced	lubricant

Forces and Motion Bingo

fulcrum	matter	wheel and axle	resistance	motion
gear	screw	lever(s)	balanced	load
momentum	inertia		compound machine	energy
efficiency (efficient)	kinetic energy	machine	Galileo	scientific method
gravity	weight	electro-magnetism	wedge	simple machines

Forces and Motion Bingo: Card No. 9

Forces and Motion Bingo

velocity	wedge	float	Sir Isaac Newton	potential energy
motion	efficiency (efficient)	joule	compound machine	load
pulley	push		work	power
machine	law	gear	weight	momentum
free fall	kinetic energy	magnets	lubricant	lever(s)

Forces and Motion Bingo

convert(ed)	push	balanced	gear	lubricant
energy	momentum	Galileo	fulcrum	joule
mass	screw		magnets	wheel and axle
free fall	resistance	weight	force	wedge
electro-magnetism	friction	inertia	newton	matter

Forces and Motion Bingo

matter	gravity	momentum	work	fulcrum
wheel and axle	tension	screw	newton	mass
inertia	scientific method		motion	Sir Isaac Newton
friction	simple machines	load	wedge	compound machine
push	energy	pulley	electro-magnetism	efficiency (efficient)

Forces and Motion Bingo

free fall	simple machines	convert(ed)	momentum	motion
screw	energy	pulley	fulcrum	power
work	float		wheel and axle	scientific method
lever(s)	weight	efficiency (efficient)	push	wedge
force	law	newton	inertia	Galileo

Forces and Motion Bingo

friction	screw	balanced	fulcrum	free fall
efficiency (efficient)	inertia	momentum	compound machine	power
gear	work		magnets	float
law	weight	pulley	electro-magnetism	convert(ed)
force	Sir Isaac Newton	push	lubricant	lever(s)

Forces and Motion Bingo

Galileo	fulcrum	balanced	matter	load
convert(ed)	potential energy	joule	screw	gear
motion	inertia		resistance	work
force	momentum	energy	weight	free fall
lubricant	gravity	newton	magnets	wheel and axle

Forces and Motion Bingo

float	suction	energy	potential energy	pull
Sir Isaac Newton	push	scientific method	mass	velocity
free fall	simple machines		motion	wheel and axle
physics	gravity	force	Galileo	wedge
gear	momentum	newton	efficiency (efficient)	power

Forces and Motion Bingo

free fall	speed	hypothesis	momentum	force
Galileo	gear	weight	work	scientific method
fulcrum	velocity		suction	energy
kinetic energy	lubricant	lever(s)	balanced	power
machine	electro-magnetism	matter	wedge	simple machines

Forces and Motion Bingo

load	pulley	efficiency (efficient)	gear	Sir Isaac Newton
kinetic energy	free fall	balanced	motion	electro-magnetism
fulcrum	power		hypothesis	potential energy
push	joule	weight	velocity	magnets
suction	momentum	machine	speed	convert(ed)

Forces and Motion Bingo

motion	convert(ed)	momentum	energy	push
Galileo	friction	potential energy	matter	velocity
speed	work		compound machine	balanced
magnets	suction	machine	gravity	hypothesis
resistance	pull	lubricant	lever(s)	newton

© Barbara M. Peller

Forces and Motion Bingo

push	speed	velocity	momentum	compound machine
float	wheel and axle	mass	machine	Sir Isaac Newton
simple machines	scientific method		physics	hypothesis
kinetic energy	lever(s)	acceleration	gravity	suction
inclined plane	tension	pull	wedge	joule

© Barbara M. Peller

Forces and Motion Bingo

convert(ed)	kinetic energy	mass	momentum	law
simple machines	hypothesis	efficiency (efficient)	energy	inertia
power	lubricant		speed	balanced
machine	matter	suction	Galileo	lever(s)
physics	pull	newton	free fall	gravity

© Barbara M. Peller

Forces and Motion Bingo

resistance	magnets	hypothesis	screw	free fall
Sir Isaac Newton	velocity	load	energy	compound machine
efficiency (efficient)	work		inertia	scientific method
suction	kinetic energy	gravity	joule	friction
pull	electro-magnetism	speed	power	mass

Forces and Motion Bingo

float	speed	matter	screw	newton
convert(ed)	push	lubricant	Galileo	joule
magnets	free fall		acceleration	inertia
power	tension	suction	electro-magnetism	gravity
law	lever(s)	pull	machine	hypothesis

Forces and Motion Bingo

float	push	friction	speed	energy
motion	newton	mass	Sir Isaac Newton	inertia
scientific method	potential energy		free fall	power
law	acceleration	suction	electro-magnetism	simple machines
inclined plane	physics	pull	velocity	tension

Forces and Motion Bingo

physics	mass	speed	balanced	hypothesis
joule	law	Galileo	float	compound machine
simple machines	energy		acceleration	suction
potential energy	kinetic energy	tension	pull	velocity
newton	friction	efficiency (efficient)	gear	inclined plane

Forces and Motion Bingo

hypothesis	speed	acceleration	Sir Isaac Newton	potential energy
magnets	work	energy	push	float
law	machine		velocity	physics
free fall	screw	kinetic energy	pull	suction
scientific method	gear	balanced	tension	inclined plane

Forces and Motion Bingo

acceleration	efficiency (efficient)	speed	push	wheel and axle
law	magnets	Galileo	suction	compound machine
weight	tension		pull	physics
potential energy	convert(ed)	inclined plane	mass	joule
free fall	velocity	hypothesis	resistance	scientific method

Forces and Motion Bingo

motion	pulley	wedge	speed	efficiency (efficient)
wheel and axle	hypothesis	acceleration	machine	velocity
tension	power		potential energy	Sir Isaac Newton
scientific method	resistance	lubricant	pull	suction
screw	fulcrum	free fall	inclined plane	law

Forces and Motion Bingo

hypothesis	pulley	potential energy	Galileo	fulcrum
law	machine	mass	scientific method	resistance
simple machines	speed		compound machine	acceleration
wheel and axle	kinetic energy	load	pull	suction
float	energy	inclined plane	convert(ed)	tension

Forces and Motion Bingo

friction	speed	Sir Isaac Newton	fulcrum	suction
joule	potential energy	magnets	velocity	compound machine
inclined plane	tension		scientific method	mass
law	convert(ed)	hypothesis	pull	acceleration
kinetic energy	matter	electro-magnetism	pulley	load

www.ingramcontent.com/pod-product-compliance
Lightning Source LLC
Chambersburg PA
CBHW051419200326
41520CB00023B/7289